Reinforced Library Binding

MARY HOFF got her undergraduate degree in biology with an emphasis on environmental issues and has a master's degree in science communication. Hoff divides her energies between caring for her husband and three children in Stillwater, Minnesota, and writing on a variety of science-related subjects.

MARY M. RODGERS became interested in global environmental concerns through her work on a country-by-country geography series for children. This encouraged her to put together Our Endangered Planet so that young readers could learn about these vital worldwide issues in a supportive and entertaining way.

Other books you'll want to read in the Our Endangered Planet series:

ANTARCTICA
OCEANS
GROUNDWATER
LIFE ON LAND
RIVERS AND LAKES
POPULATION GROWTH
TROPICAL RAIN FORESTS
SOIL

Our Endangered Planet
ATMOSPHERE

Mary Hoff
and
Mary M. Rodgers

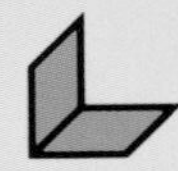

LERNER PUBLICATIONS COMPANY • MINNEAPOLIS

Thanks to Dr. Dean Abrahamson, Erica Ackerberg, Zachary Marell, and Gary Hansen for their help in preparing this book.

Words in **bold** type are listed in a glossary that starts on page 67.

LIBRARY OF CONGRESS CATALOGING-IN-PUBLICATION DATA

Hoff, Mary King
Our endangered planet. Atmosphere / Mary Hoff and Mary M. Rodgers.
p. cm.
Includes index.
ISBN 0-8225-2509-7 (lib. bdg.)
1. Atmosphere—Juvenile literature. 2. Ozone layer—Juvenile literature. 3. Greenhouse effect, Atmospheric—Juvenile literature. [1. Atmosphere. 2. Ozone layer. 3. Global warming. 4. Greenhouse effect, Atmospheric.] I. Rodgers, Mary M. (Mary Madeline), 1954- . II. Title.
QC863.5.H64 1995
363.73'87—dc20 94-30899
CIP
AC

Manufactured in the United States of America

1 2 3 4 5 6 - I/JR - 00 99 98 97 96 95

Front cover: *Satellites measure levels of an atmospheric gas called ozone and compare the findings to reveal monthly changes over the years. These images show ozone amounts over the continent of Antarctica between 1979 and 1990.* ***Back cover:*** *(Left) Drivers in Belarus, a newly independent nation in eastern Europe, line up to buy gas. Fuel-burning cars add tons of carbon dioxide to the air and contribute to rises in worldwide temperatures. (Right) A police officer in California uses a bicycle, instead of a gas-powered squad car, to cover his beat.*

Recycled paper

This product contains 50 percent recycled paper, 10 percent of which comes from postconsumer recycled paper.

Recyclable

CONTENTS

OUR ENDANGERED PLANET

In the 1960s, astronauts first traveled beyond the earth's protective atmosphere and were able to look back at our planet. What they saw was a beautiful globe, turning slowly in space. That image reminds us that our home planet has limits, for we know of no other place that can support life.

The various parts of our natural environment—including air, water, soil, plants, and animals—are partners in making our planet a good place to live. If we endanger one element, the other partners are badly affected, too.

People throughout the world are working to protect and heal the earth's environment. They recognize that making nature our ally and not our victim is the way to shape a common future. Because we have only one planet to share, its health and survival mean that we all can live.

High above our heads, atmospheric gases make it possible for life to exist on the earth. Some of these gases let in heat and sunlight, which all living things need to survive. Other gases block the sun's harmful rays.

But some of our actions are changing or destroying the kinds and amounts of gases in the atmosphere. We hope that by learning how these gases affect our planet, we can teach others. By seeing how we are harming the atmosphere, we can change our behavior. And by working to protect the atmosphere, we can ensure that life on our planet will continue well beyond the twenty-first century.

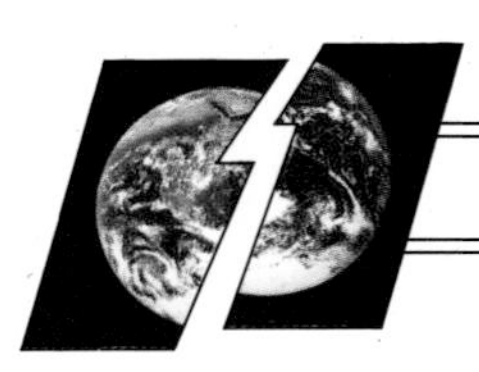

CHAPTER ONE

THE EARTH'S QUILT

If you live in a part of the world where it gets cold in the winter, you probably know how nice it feels to snuggle under the thick layers of a padded quilt. But no matter where you live, you are always surrounded by a sort of planetary quilt. It's the blanket of air called the atmosphere. Like any good quilt, the atmosphere covers and warms you. But rather than safeguarding just one person, the atmosphere protects everything on our planet.

IT'S A GAS!

The atmosphere that blankets our world is sometimes simply called air—a mixture of gases that stretches over land, over water, and far above the earth's surface. We can't usually see these gases. When the wind hits our face, though, we're feeling moving air.

(Left) ***Gases in our atmosphere form protective layers that enable us to survive even in cold weather.*** (Above) ***Swirling air, although not visible in itself, makes this flag snap and flutter.***

Carbon Dioxide Molecule (CO_2)

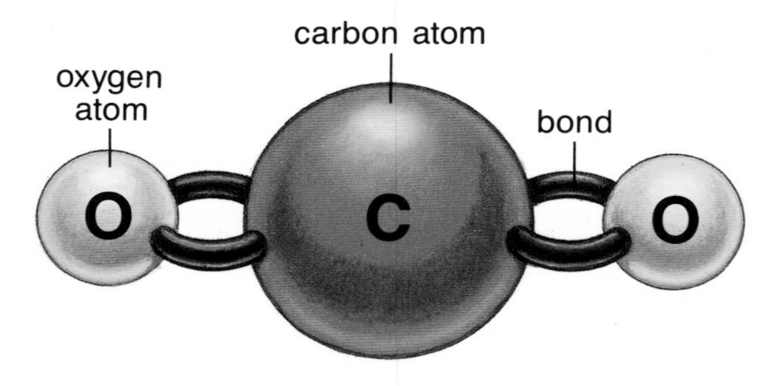

A Cupful of Air

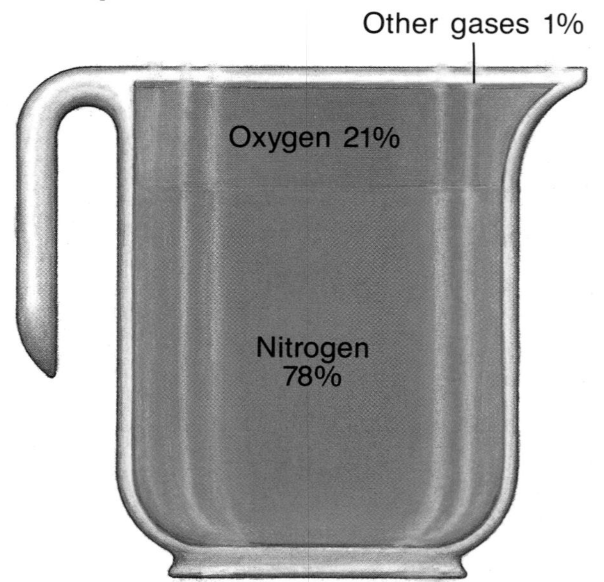

When we breathe, we're taking air into our lungs. Most of the sounds we hear are carried by waves of air.

Each of the gases in the air is made up of specific **atoms** (invisible bits of matter), which combine to form **molecules.** Near the earth, there are usually lots of gas molecules, and the pulling force of gravity keeps them very close together. Way above our planet, where the pull of gravity is weaker, there are fewer molecules, and they're farther apart. When the gas molecules are close together, the atmosphere is "thick." When they're farther apart, the atmosphere is "thin."

If we mixed up a cupful of air, we'd fill just over three-quarters of the cup (78 percent) with nitrogen gas molecules. Next, we'd add oxygen almost all the way to the brim. This gas makes up nearly 21 percent of the atmosphere. At the very top of the cup, we'd still have room to add a drop of argon, **carbon dioxide,** helium, water vapor, neon, and **ozone.** These and several other gases together form only about 1 percent of all the gases in the air.

DRILL FOR IT!

As they fall, snowflakes pick up gas molecules that are in the air. In frozen areas, such as the North and South Poles, the ice and snow never melt, and each new deposit of snow buries the previous layer. As a result, the icy layers trap the molecules, sometimes for thousands of years. To reach this museum of atmospheric history, scientists use special tools to drill deeply into the ancient ice. The drills take pieces of ice—called **core samples**—from each layer. Scientists then study the samples in search of possible climatic patterns.

Among the most revealing core samples are those from Antarctica, an icebound continent surrounding the South Pole. From the samples, scientists have found that during the past 160,000 years the amount of CO_2 in the atmosphere over Antarctica has corresponded to the pattern of local air temperatures. When CO_2 levels went up, so did the air temperature. But when temperatures cooled, CO_2 levels dropped. From such long-term patterns, researchers hope to predict future atmospheric conditions.

A scientist in Antarctica holds up a small core sample he has just taken.

The atmosphere's gases work together to benefit all living things on our planet. They keep the earth warm—but not too warm—by trapping some of the sun's blazing heat. The atmosphere also lets in just enough sunlight to help plants grow. Some atmospheric gases shield us from very harmful rays that come from the sun.

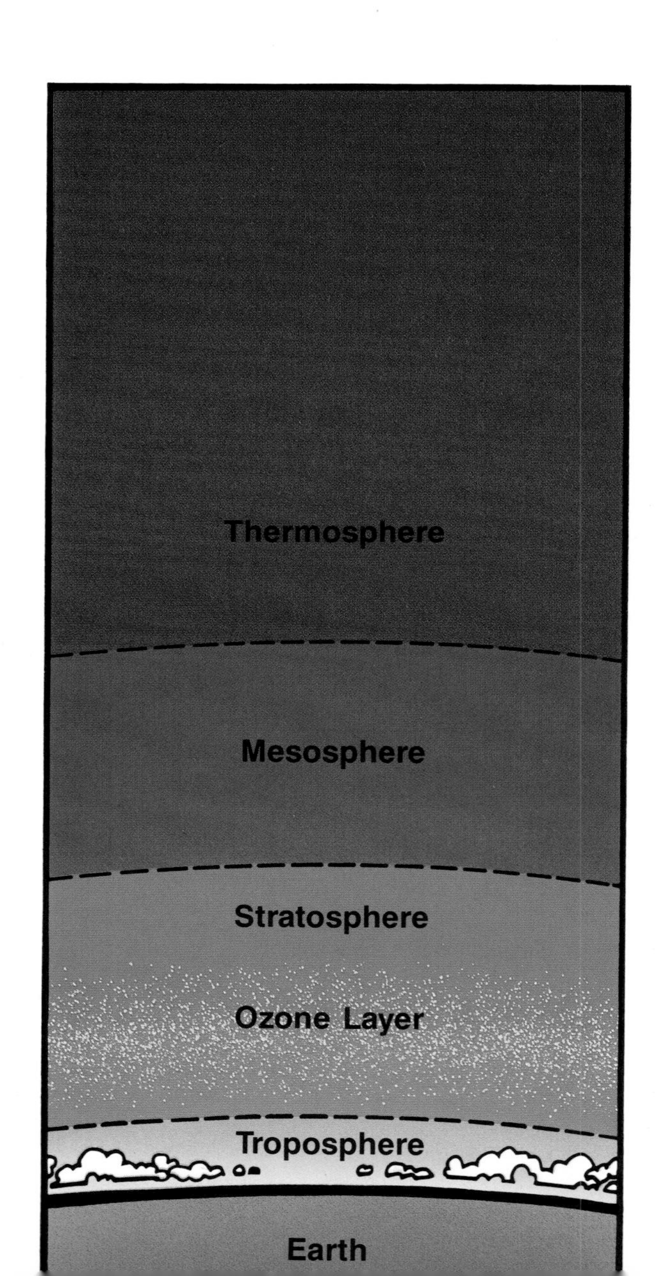

UP, UP, AND AWAY

Even though we experience the atmosphere close to home, this protective covering stretches far above our planet in four separate layers. The **troposphere,** the layer nearest the earth, is what we usually mean when we use the word "sky." Reaching up about 7 miles (11 kilometers), the troposphere holds the oxygen we breathe. It's where most clouds, rain, and snow form. This layer also contains most of the world's **air pollution.**

Starting about 7 miles (11 kilometers) above the earth's surface is the **stratosphere.** Pilots like to fly jets in the stratosphere because it has few clouds or dangerous storms. The upper part of this layer—located nearly 30 miles (48 kilometers) above the earth—has most of our atmosphere's ozone.

Next comes the **mesosphere.** This layer stretches from about 30 miles (48 kilometers) all the way to 50 miles (80 kilometers) above our planet. Conditions in the topmost section of the mesosphere are fierce. Temperatures can drop to –165 °F (–110° C), and very strong winds are common.

These cirrus clouds are floating in the troposphere, the layer of the atmosphere closest to earth.

On long journeys, jet planes cruise above the clouds in the stratosphere.

In March of 1993, northern Georgia was surprised by at least 16 inches (41 centimeters) of snow.

The fourth layer—the uppermost layer of our planet's atmosphere—is called the **thermosphere.** It's hard to tell exactly where this layer ends and where outer space begins. Most scientists put the thermosphere's outer edge at about 600 miles (965 kilometers) above our planet. The air in the thermosphere is very thin, because so few gas molecules exist that high up. As a result, the thermosphere has little protection from the scorching heat of the sun. Temperatures here can top 2,000° F (1,102° C)!

WEATHER AND CLIMATE

The troposphere is the stage for two dramatic actors—weather and climate. Though different, each player has a starring role in the story of atmosphere.

Take a look outside right now. What's it like? Maybe rain is splashing against the windowpane. Perhaps there's a pile of fresh snow waiting to be turned into snowballs. Or maybe it's a boiling hot day, and you can't wait to get to the beach.

Your answer to ''What's it like?'' describes the weather—the daily changes in the atmosphere. Weather differs a lot from place to place and can even vary in the span of a few hours. If you've ever dressed for a sunny day and then been caught in an unexpected rainstorm, you know what a quick-change artist the weather can be.

Seasonal rainfall lashes a shop in the Central American nation of Guatemala. The country lies in a tropical area where rain is heavy from May through October every year.

What's the weather like in your area over the course of a year? To answer that question, you'd need to keep track of the weather from day to day, during all seasons, and for many years. Your records would tell you that, even though weather varies a lot, in a certain area it forms a pattern. That pattern—what the weather tends to be like over a long period of time—is known as the climate.

It's important to remember that climate is only a pattern. Even the coldest climate

Average Temperatures/Rainfall in Selected Mexican Resort Areas

Sun-seeking vacactioners often look at average monthly temperatures and rainfall levels to find out the best time to travel. The figures below are for resorts in Mexico from January through December. Acapulco lies on the country's Pacific coast. The combined region of Cancun/ Cozumel faces the Caribbean Sea.

	J	F	M	A	M	J	J	A	S	O	N	D
ACAPULCO												
temperature (°F)	78	78	79	80	83	83	83	83	82	82	81	79
temperature (°C)	26	26	26	27	28	28	28	28	28	28	27	26
rainfall (in)	0.4	0.03	0.0	0.01	1	17	9	10	14	7	1	0.5
rainfall (cm)	1	0.08	0.0	0.03	3	43	23	25	36	18	3	1
CANCUN/COZUMEL												
temperature (°F)	73	74	76	79	80	80	81	81	80	79	76	74
temperature (°C)	23	23	24	26	27	27	27	27	27	26	24	23
rainfall (in)	3	1	2	2	7	7	3	5	8	12	5	3
rainfall (cm)	8	3	5	5	18	18	8	13	20	30	13	8

(Source: *World Climatic Data*. Length of study: 15 years)

Knowing the region's climate allows these farmers in Thailand, Southeast Asia, to plant their rice crop to yield the best harvest.

will have some hot days, and rain falls in the driest desert, at least sometimes. But over years and years, these unusual days don't really interrupt the climatic pattern very much.

In fact, knowing a climate's pattern helps people plan ahead. Farmers in Thailand, Southeast Asia, for example, can usually count on rain-bearing winds to water their rice crops during certain months. Bridge builders in eastern Russia know they need to design their structures to withstand extremely cold winter temperatures. Resort owners in Mexico can expect that each February, winter-weary tourists from the United States and Canada will flock to Mexico's sun-drenched beaches. And the visitors can trust that local temperatures will be warmer than those back home.

Although there have always been summer frosts and winter heat waves, the world's climate has been fairly predictable over many centuries. In fact, we've come to count on the climate to stay the same. If it were to change a lot, we'd have to change a lot, too.

SCORE:
UV-BEES 247
OZORS 401
POLLUTZ 262

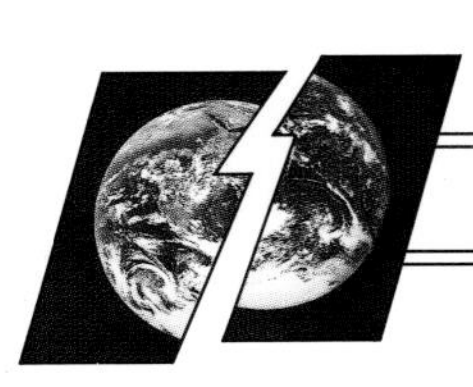

CHAPTER TWO

A Hole in Our Defenses

Imagine a video game with this story line. The earth is being bombarded by deadly rays from a powerful source far out in space. Fortunately for earthlings, tiny devices in the atmosphere pick off the rays before they can strike the earth and harm its plants, animals, and people. What a relief!

But wait. Suddenly the earthlings realize that something or someone is destroying these lifesaving devices. As it turns out, the evildoers are not from outer space but are the very people the devices are meant to protect. Now what do the earthlings do?

(Left) An imaginary video game shows ozone gas molecules keeping out the sun's ultraviolet B (UV-B) rays. Ozone forms a thin layer to shield our planet from UV-B, which is harmful to humans and other living things.

Although this video game probably doesn't exist, the story line is real. Deadly **ultraviolet B (UV-B) rays,** a form of light from the sun, are reaching the earth right now. The tiny devices in the story line are ozone gas molecules in the upper stratosphere. Ozone floats in a very thin sheet called the **ozone layer,** absorbing most of the UV-B before it reaches us.

The sun naturally gives off UV-B, along with the warmth and sunlight living things need to thrive. But too much UV-B can damage our skin and eyes, can make it harder for our bodies to fight disease, and can slow plant growth. Without the fragile ozone layer, the living things that share this planet could not survive.

NO ZONE LIKE OZONE

A molecule of ozone is made of three atoms of oxygen. Scientists refer to ozone gas by its symbol—O_3. Oxygen gas, or O_2, consists of two atoms of oxygen tightly bound together. Both gases are naturally found in the ozone layer.

Even though these two gases are made of oxygen atoms, oxygen and ozone are vastly different. For example, we need oxy-

This two-part illustration depicts how incoming UV-B helps ozone (O_3) to form in the stratosphere. In the left image, UV-B (purple arrow) breaks an oxygen molecule (O_2) into two single oxygen atoms, which combine with unbroken O_2 molecules to create O_3. In the right image, incoming UV-B is splitting an O_3 molecule into O_2 and a single atom of oxygen. Again, the single atom joins unbroken O_2 to make more O_3.

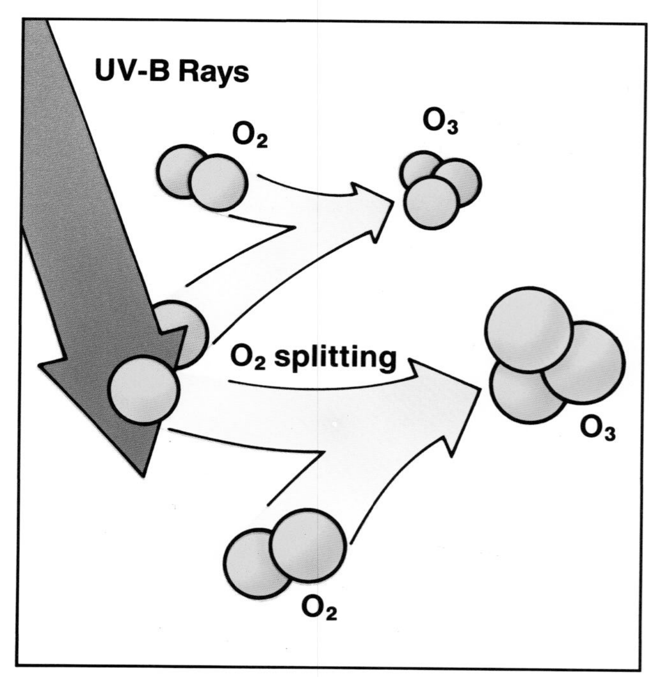

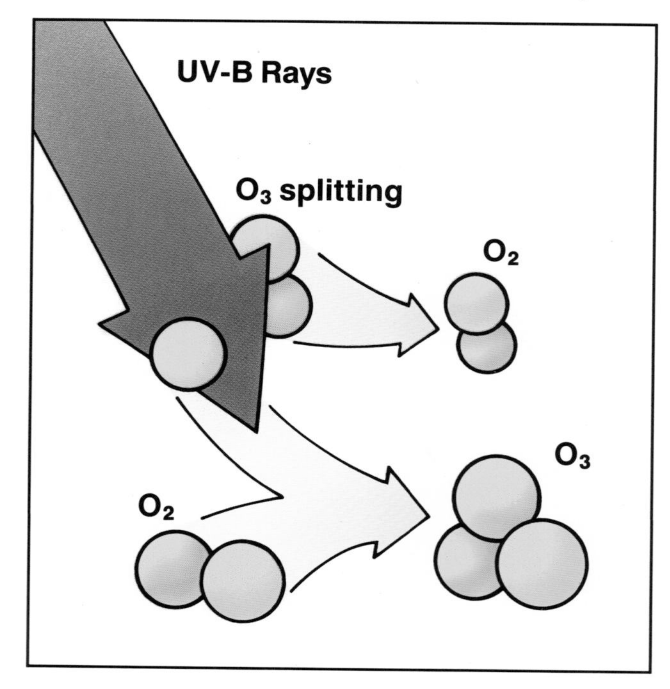

gen to breathe, but breathing too much ozone could kill us. When ozone is floating high in the stratosphere, the smelly, blue gas makes life on our planet possible by shielding us from UV-B. But when ozone gathers in the troposphere, the gas is a form of air pollution.

In the stratosphere, ozone gas molecules come together, break up, and reform in a continuing cycle. Here's how the stratospheric cycle works. Powerful rays from the sun strike O_2 molecules, undoing the bond between the two atoms. The free, single atoms of oxygen hook onto nearby O_2 molecules and form O_3. At the same time, the sun's rays are hitting existing O_3 molecules, which break up into single atom oxygen and O_2 molecules. These recombine to create new ozone molecules. This process, which occurs over and over again, maintains the ozone layer.

ENTER THE VILLAINS

Now that we know how important ozone is to us, let's face the downside of the story.

When they were first invented, aerosol spray cans and plastic foam packaging both contained chloro-fluoro-carbons (CFCs)—chemicals that are harmful to ozone.

For many years, people have been making and using certain chemicals that actually attack the ozone layer. The gases in these products rise through the troposphere to the stratosphere and destroy O_3 molecules.

Even though O_3 molecules can replace themselves, the gases have been attacking at breakneck speed. Because O_3 molecules can't replace themselves fast enough, the ozone layer is thinning. In some places, it has become so thin that scientists speak of "holes" in the layer.

***CFCs became key parts of many manufacturing steps, such as cleaning high-tech equipment** (left). **Another chemical that destroys ozone is halon, which is a part of gas fire extinguishers** (right).*

The most important ozone wrecker is a compound, or group of chemicals, called **chloro-fluoro-carbons (CFCs).** Chemists who wanted to find a way to keep refrigerators cool without ice invented CFCs in 1928. Other scientists soon found many other jobs for this remarkable compound. By the 1970s, CFCs were a key part of auto air conditioners and were helping to push out the liquid in spray cans. CFCs were in plastic foam packaging, as well as in cleaners for electronic equipment and surgical instruments.

But CFCs are not the only ozone-destroying chemicals. Halons, methyl bromide, carbon tetrachloride, and methyl chloroform are also harmful to ozone. Halons are found mainly in fire-fighting equipment. Methyl bromide is one of the world's most popular pest killers. The other

two compounds show up in bug sprays, spot removers, and cleansers.

GETTING THE DRIFT

For most of our history, few people knew about or thought about CFCs and the ozone layer. In 1973, however, two U.S. scientists—Sherwood Rowland and Mario Molina—became interested in CFCs.

CFCs are very stable compounds, meaning they break down slowly and stay in the atmosphere for a long time. Rowland and Molina discovered that the gases in used CFCs slowly drift upward through the troposphere and stratosphere. There, far above the earth's surface, the same powerful rays that help ozone to form also bombard CFC molecules. During the attack, chlorine atoms—which have a strong liking for ozone—break off and roam free.

For decades, homes and businesses have leaked or released CFCs into the atmosphere. The chemicals slowly drift upward until they reach the ozone layer and begin destroying ozone molecules.

DEDICATED VOICES FOR CHANGE

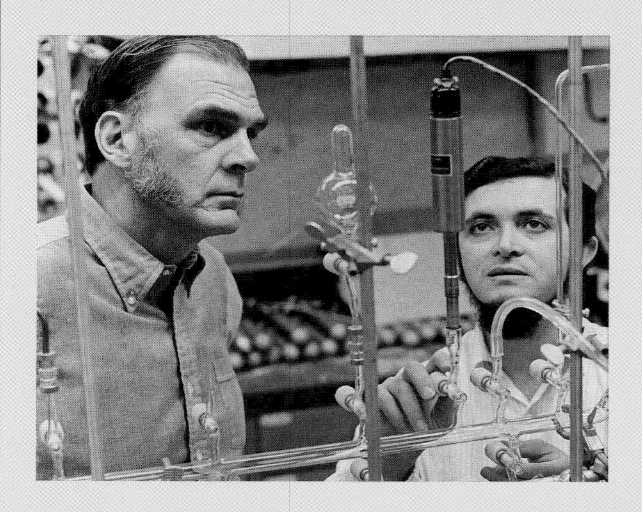

Sherwood Rowland (left) *and Mario Molina began researching CFCs on the campus of the University of California at Irvine.*

In the early 1970s, two U.S. chemistry professors—F. Sherwood Rowland and Mario J. Molina—began studying the fate of CFCs in the stratosphere. By 1974, they had gathered enough evidence to warn the scientific community that CFCs were causing serious ozone loss. They also recommended a complete ban on CFC releases into the atmosphere. But companies that made or used CFCs strongly questioned Rowland and Molina's study. These corporations hired experts to do additional research, which cast doubt on the chemists' findings. But the two scientists didn't give up.

By the early 1980s, the research of other scientists supported the 1974 conclusions. The explosive news in 1985 from British scientists in Antarctica—that declines in the ozone layer over the continent had produced an "ozone hole"—further confirmed Rowland and Molina's original work. Within a few years, CFC-making corporations were developing substitutes that did not harm ozone. With these new chemicals, manufacturers were able to begin introducing CFC-free air conditioners for cars and homes.

Rowland and Molina both continue to study issues related to ozone. Twenty years ago, neither would have predicted that their report would lead to worldwide agreement among scientists in industries, governments, and universities.

When a free chlorine atom meets an ozone molecule, it attacks the ozone like a cat pouncing on a mouse. After the dust has settled, three things have happened. A molecule of O_2 is free. A single atom of oxygen is on the loose. And a liberated chlorine atom is still looking for partners.

Did you notice that the chlorine atom came out of the battle in one piece? Because chlorine survives unharmed, just one CFC molecule can destroy thousands of ozone molecules. Later studies showed that gases from halons and methyl bromide, which both contain bromine, are also rising into the stratosphere. Heavier than chlorine, bromine atoms have an even easier time destroying ozone molecules.

GOING...GOING...

From their studies of CFCs, Rowland and Molina predicted a drop in the amount of global ozone. But, because there was so little evidence to prove their theory, other scientists were skeptical, at least until 1985.

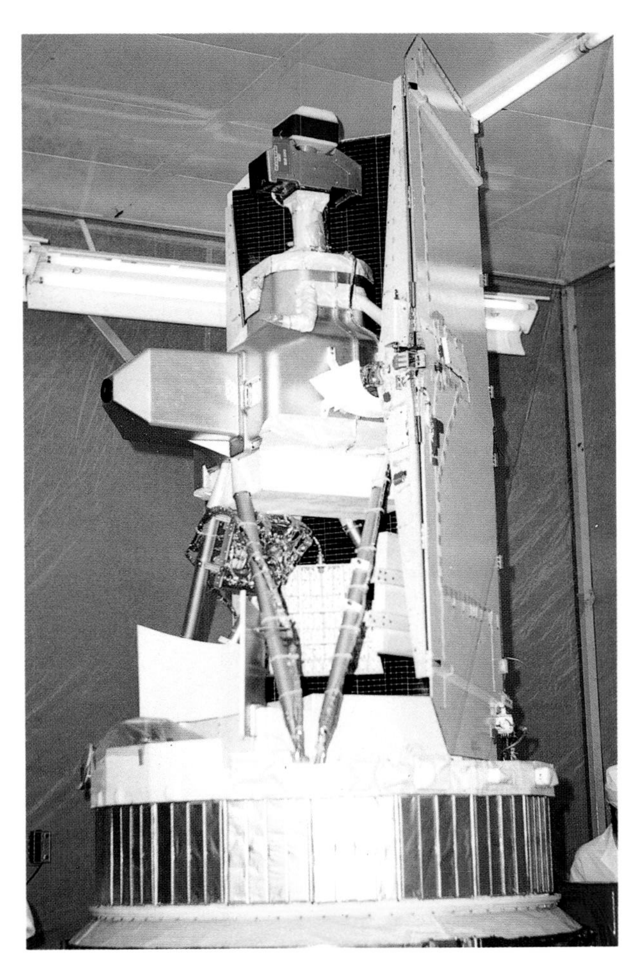

While the U.S. scientists Rowland and Molina were studying ozone, a Nimbus-7 satellite** (shown here before being launched)* ***was recording data that would eventually prove their findings.

In that year, British scientists studying the ozone layer over Antarctica—the cold continent surrounding the South Pole—reported a 40-percent drop in the area's ozone. This amount of ozone loss, the scientists said, formed a gaping hole as big as the United States.

British researchers living in the wilds of Antarctica first told the world in 1985 that the ozone layer over the continent was thinning.

Shocked by the findings, many scientists began to closely monitor the thinning of the ozone layer above Antarctica. By 1987, the hole over Antarctica was twice the size of the United States. In 1993, the ozone concentration over Antarctica dropped by a total of 56 percent.

Scientists studying ozone in other areas of the world also had bad news to report. In 1987, the ozone shield weakened over Australia. Scientists monitored a huge ozone hole that developed over the North Pole in 1988. Two years later, measurements revealed that the ozone layer covering Europe and North America had thinned by 10 percent. By 1994, satellite data were showing the lowest levels of global ozone ever recorded.

PAYING THE PRICE

So we've weakened our ozone defenses. Is that really such a big deal? Ana, a South American girl in Punta Arenas, Chile,

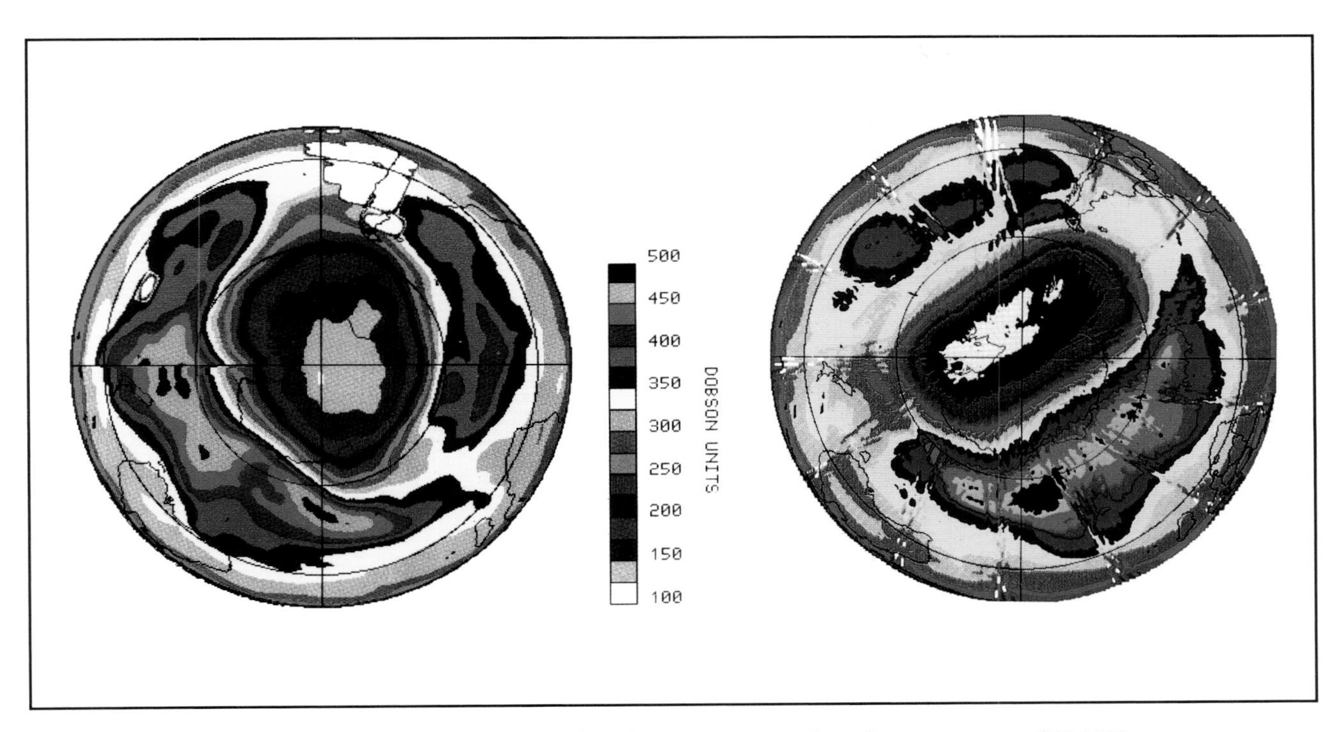

Satellites that carry an instrument called a Total Ozone Mapping Spectrometer (TOMS) measure ozone levels and help scientists monitor the ozone layer. In October 1991, the TOMS picture** (left) **revealed low readings** (in pink) **over Antarctica. A month later, another picture** (right) **indicated that ozone was also decreasing over Australia, South America, and Africa.

thinks so. Her parents won't let her play outdoors between 10:00 A.M. and 3:00 P.M. —the time when the sun's rays are strongest. Ana's parents fear that Antarctica's ozone hole is letting too much UV-B reach Punta Arenas, the city closest to the South Pole. Throughout the world, the number of people with skin cancer or an eye disease called cataracts is going up. Many scientists blame the rise on increased UV-B rays.

In Australia, a surfer puts cream on his nose to protect his skin from overexposure to UV-B.

Humans are not the only living things at risk from ozone loss. **Phytoplankton,** tiny plants that live near the ocean's surface, provide food for many other forms of water life. These plants are the basis of the ocean's **food web.** Because of increased UV-B, however, phytoplankton are growing more slowly than they have in the past. If the ozone layer continues to thin, many sea animals that depend on the plants could be threatened with starvation.

High levels of UV-B can also stunt—or even kill—fish and shellfish when they're young. The eggs of other water creatures, such as certain types of frogs, already have

Tiny phytoplankton (shown enlarged) *live in the ocean around Antarctica. Too much UV-B is causing the plants to reproduce more slowly. Ocean life that eat the plants could starve if phytoplankton amounts go down sharply.*

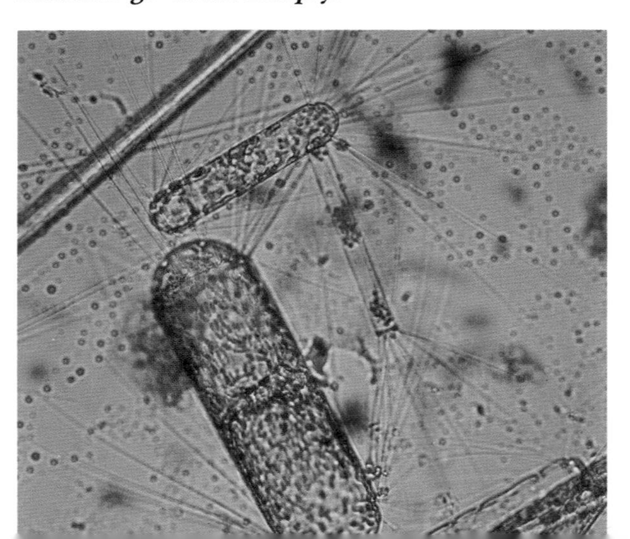

At an experimental plot in Maryland, scientists have placed UV-B lamps over a small field of soybean plants. The lamps increase the plants' exposure to UV-B and so mirror growing conditions if we continue to lose ozone.

been harmed by the increase in UV-B. Fish catches could decrease if large numbers of sea animals fail to reach full size or become too deformed to harvest.

Ozone loss may affect land-based plants and animals, too. Tests have shown that plants exposed to high levels of UV-B aren't as healthy as normal plants. They grow fewer leaves, take in less water, and are puny. Sometimes the seeds never produce plants at all.

The problems seem to be worse for crops, particularly soybeans. Scientists have found that, for some types of soybeans, the number of plants in a field drops sharply after exposure to UV-B. In Australia, fields of wheat, sorghum, and peas already have been harmed.

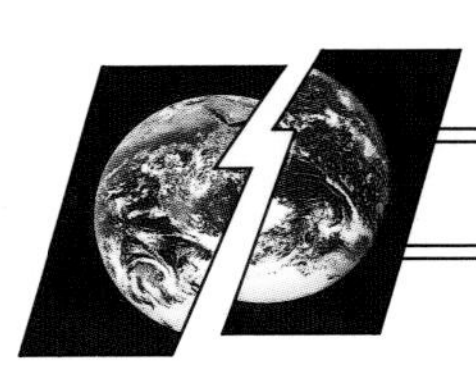

CHAPTER THREE

THE HEAT IS ON

Likening the atmosphere to a cozy quilt reminds us that the atmosphere helps keep our planet safe and warm. We've already talked about ozone's job of blocking the sun's harsh UV-B rays. Other gases in the atmosphere work like a quilt's padding to hold in heat.

GASES AT WORK

Carbon dioxide, ozone, methane, and about 30 other atmospheric gases allow the sun's light to reach the earth's surface. This light takes several forms. The ozone layer keeps out most of the UV-B. But two other forms, **infrared radiation** and **visible light,** get through. Infrared radiation is an invisible form of light. We can't see it, but we feel it as heat when we're in the sunshine. Visible light is what makes the difference between night and day and shines through in rainbows.

When visible light and infrared radiation hit the earth's surface, much of their energy is changed into heat. The surface sends this warmth back toward outer space. As the heat floats upward through the troposphere, it meets atmospheric gases. Some of the heat escapes to outer space. But the gases also block a lot of the heat. This heat becomes trapped and keeps our planet warm.

(Left) After a heavy rainfall, rays from the sun pass through the drops of water that remain in the air. We "see" the sunlight in rainbows.

Gases that trap heat in this way are called **greenhouse gases,** and their warming ability is known as the **greenhouse effect.** The names come from a garden building—a greenhouse—that is made mostly of glass. In the same way that a glass greenhouse lets in sunlight and traps warmth, atmospheric gases block heat but

One way to understand how greenhouse gases work in the atmosphere is to walk through a glassed-in greenhouse here on earth. This building lets in sunlight and traps heat, both of which are needed to nourish all kinds of plants. This particular greenhouse is on Iceland, a very cold island nation between the North Atlantic and Arctic Oceans.

The greenhouse gas carbon dioxide (CO_2) is a partner in photosynthesis (green arrows) and respiration (red arrows). Using energy from the sun, plants take in CO_2 and water and release oxygen to make their own food. When once-living things decompose (break down) during respiration, they take in oxygen and release CO_2.

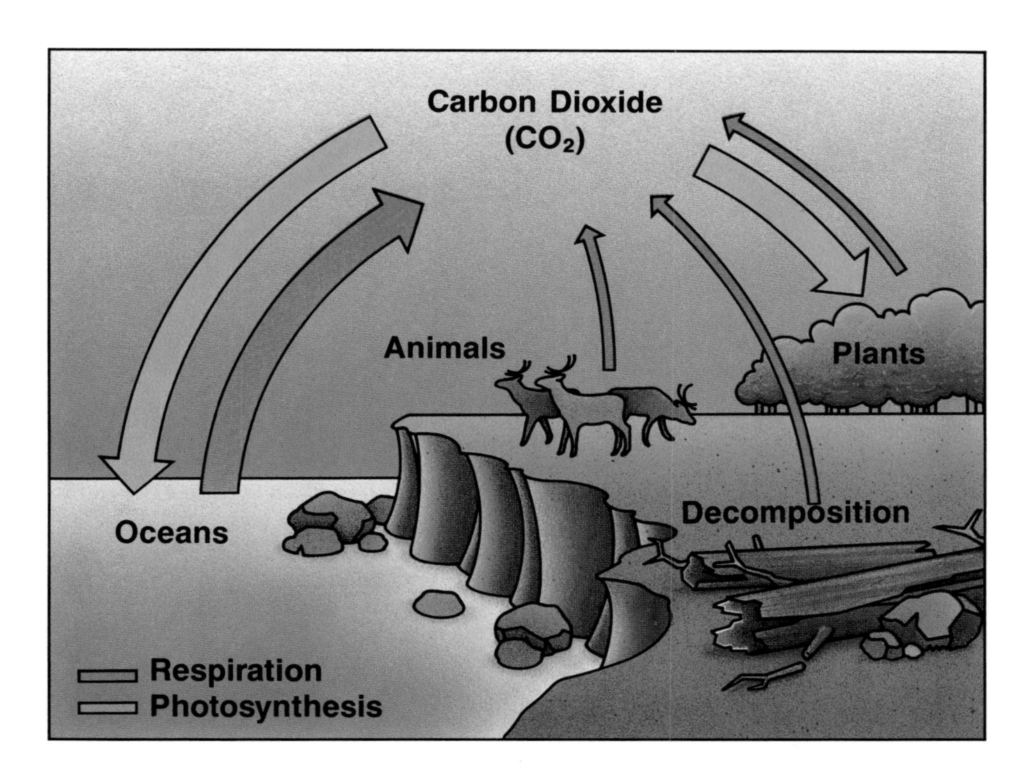

not light. The plants inside a greenhouse can grow even if the weather is very cold outside. Likewise, living things on our planet thrive even though conditions beyond our atmosphere cannot support life.

Greenhouse gases are part of our planet's natural makeup. The most important greenhouse gas is carbon dioxide, or CO_2. Although this gas makes up only 0.035 percent of our atmosphere, CO_2 is a key part of **photosynthesis,** the process by which plants make food. Methane and nitrous oxide are other natural greenhouse gases. Cows belch methane gas as they digest food. The plowing of fields releases nitrous oxide from the soil into the air.

ADDING OUR BIT

If you've ever been wrapped in a warm blanket on a hot night, you know that there can be too much of a good thing. Suddenly the blanket isn't so comfortable because it's trapping too much heat. Some scientists believe that our atmospheric quilt is doing the same thing. These experts think a worldwide increase in greenhouse gases is raising global temperatures—a trend they term **global warming.**

In some ways, our actions are adding greenhouse gases to the atmosphere. The amount of CO_2 goes up when we burn **fossil fuels**—the gasoline that runs our cars, the coal that fuels our factories, the natural gas that heats our homes. Every year, throughout the world, humans add more than 20 billion tons (18 billion metric tons) of CO_2 to the atmosphere just from burning fossil fuels.

Some of our actions—such as burning fossil fuels to run factories and cars—are adding more CO_2 to the atmosphere. Scientists believe that the increases are helping to raise global temperatures.

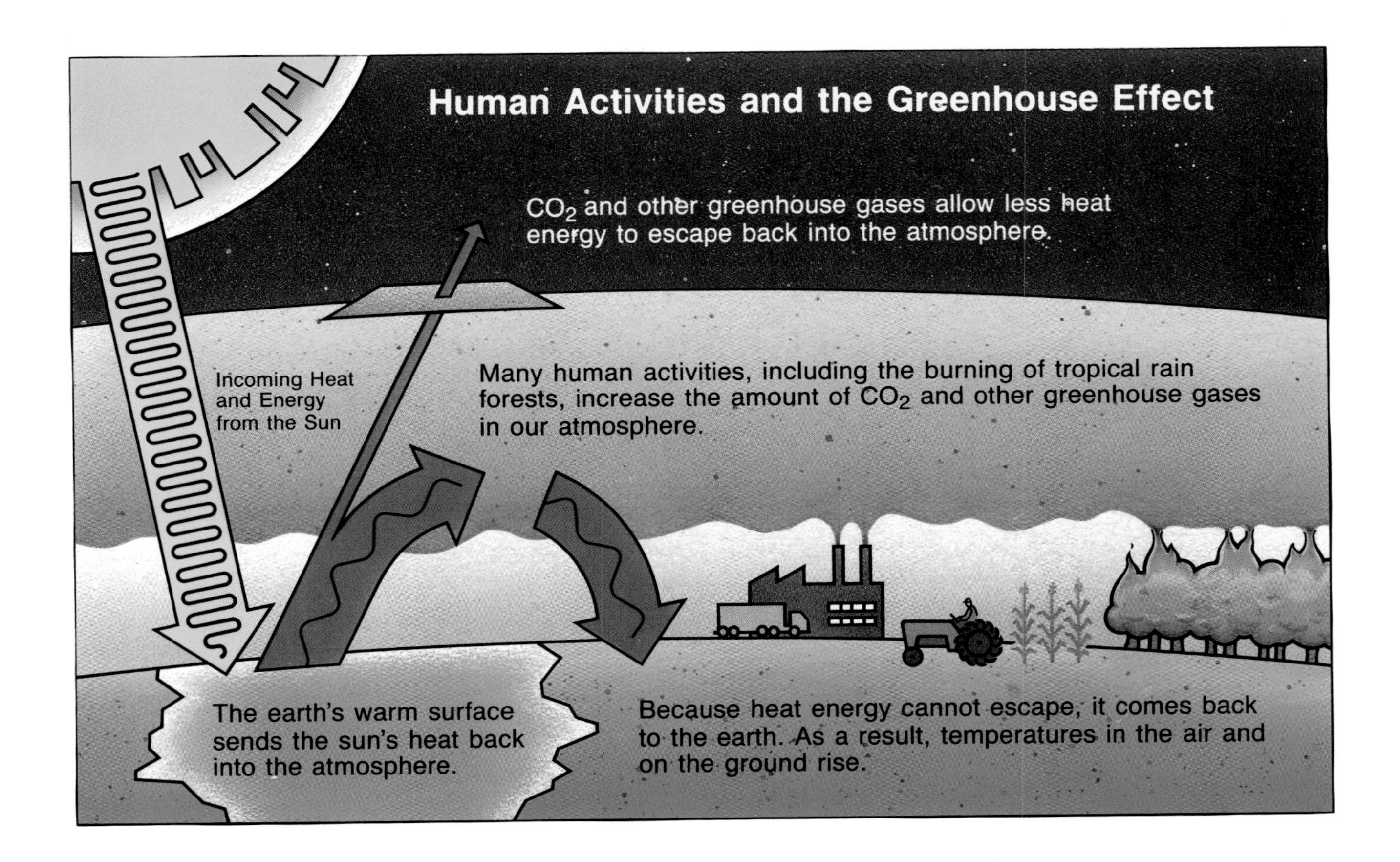

We put in at least another 3.3 billion tons (3 billion metric tons) of CO_2 each year through deforestation, the large-scale cutting and burning of trees. People throughout the world uproot and then burn trees and other plants to create farmland or to open up sites for cities, mines, and factories. Deforestation accounts for about 20 percent of all the CO_2 we add to the air.

Atmospheric methane is also on the rise. In recent years, the amount of this greenhouse gas has increased annually by around 4 million tons (3.6 million metric tons). What accounts for the increase? The biggest contributors are **anaerobic bacteria**—tiny life-forms that break down dead matter and give off methane in the absence of oxygen.

Cows release methane as the bacteria in their stomachs digest the grass the animals have eaten. Anaerobic bacteria in soggy soils—such as rice fields, bogs, and swamps—also produce methane in great amounts. In landfills (garbage dumps covered with soil), anaerobic bacteria work to break down what we've thrown away. In the process, the bacteria manufacture methane.

As they digest the grass they've eaten, cattle release methane by belching this greenhouse gas, sometimes as often as twice a minute. Every year, cattle throughout the world add nearly 100 million tons (91 million metric tons) of methane to the atmosphere.

Food that enters a cow's stomach goes through a complex process before being fully digested. Steps during this process make methane. The red line follows the path the grass takes after it's been swallowed. The food goes down the food pipe to storage areas in the stomach where the food is softened by tiny anaerobic bacteria. Eventually, the food goes back up the food pipe as cud, which the animal rechews and reswallows. As the blue line shows, the bacteria further break down the cud, which is then sent to other parts of the stomach to finish the digestive process.

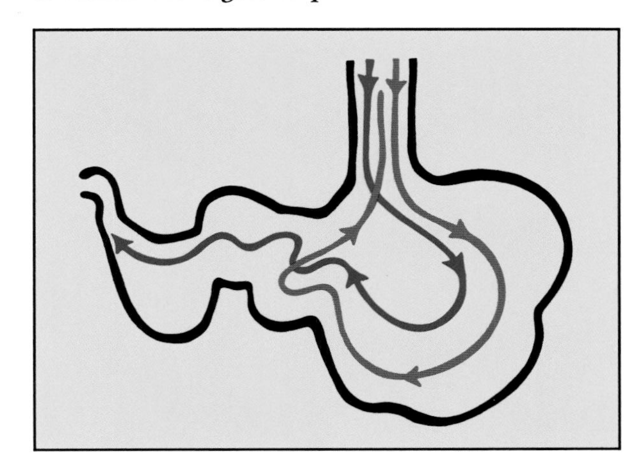

A farmer in the Philippine Islands of Southeast Asia carefully sets a rice seedling in a paddy (wet field). Methane gas, which exists in soggy soils, can travel upward through the long, hollow stems of rice plants and out into the atmosphere.

We increase the amount of methane in various ways. When we build more landfills to bury garbage, for example, we are creating a new place for methane to be formed. To help feed the world's growing population, we plant more rice and raise more cows. But in taking these actions to meet our planet's immediate needs, we also raise the amount of methane in the atmosphere.

TERMITES ON TRIAL

Methane, a natural greenhouse gas, traps even more heat than CO_2 does. The amount of atmospheric methane has gone up dramatically in the past 200 years. As a result, scientists are closely studying various methane-producing environments, such as dairy farms, wetlands, and rice paddies.

Some researchers are also looking at termites. These small insects feed on wood, grass, or soil. As they digest, termites—especially those that eat soil and live in tropical areas—release methane. But just how much methane do they give off?

In the early 1980s, lab scientists estimated that termites accounted for 30 percent of all the methane added to the atmosphere each year. A U.S.-Australian report suggested in 1990 that these early estimates were wrong and that the annual amount was actually about 2 percent.

Researchers in Europe raised the issue again in the mid-1990s. They said that termites aren't the only insects that release lots of methane. So do certain other kinds of tropical bugs, mainly cockroaches, millipedes, and scarab beetles. The estimates from this report varied from 2 percent to as much as 60 percent per year. More studies are needed, though, to find out if termites and other insects are really the guilty ones.

Termites in Malaysia's tropical rain forest eat soil and wood.

Amounts of another greenhouse gas—nitrous oxide (N_20)—are going up more slowly. Like CO_2, N_20 comes from burning fossil fuels, mainly coal. Like methane, N_20 is also linked to farming practices. When farmers fertilize their soil to help plants grow, they often use chemicals that contain nitrogen (the "N" of N_20). The soil takes in these nitrogen-rich fertilizers and then releases N_20 into the air through plowing.

Nitrous oxide is added to the troposphere each year in much smaller amounts than are methane or CO_2. But N_20 steadily builds up. It can remain in the troposphere for at least 150 years before finally rising to the stratosphere and breaking down into harmless molecules. CO_2 also has a long lifetime in the atmosphere, as long as 200 years. Methane, on the other hand, isn't a very stable gas and breaks down within 10 years.

Using a modern tractor, a farmer adds liquid nitrogen fertilizer to his cornfield. Nitrogen is part of nitrous oxide, another natural greenhouse gas whose amount is rising because of human actions.

CFCs are greenhouse gases that only people have produced, mostly in the twentieth century for industrial purposes. They are not naturally found on our planet. Yet, these artificial gases also contribute to global warming and remain in the atmosphere for many decades.

NOT SO SIMPLE

We know that greenhouse gases block some heat from leaving the atmosphere and that this action helps to warm the earth. We know that the amount of these gases in the atmosphere is rising. Can we then say that the earth's temperature is going up, too? Although there does seem to be a pattern of global warming, scientists are unsure about how the earth will react over time.

Some climate experts, for instance, believe air, land, water, and living things react to increases in greenhouse gases in a way that works *against* the predicted change. They call this natural response **negative feedback.** Here's an example. If greenhouse gases do raise the earth's temperature, the increased warmth might cause more clouds to form. But some types of clouds reflect sunlight back toward outer space, an action that might result in a cooling of the earth.

Could negative feedback work against global warming? Maybe. But scientists are also talking about **positive feedback.** This is a process that speeds up, rather than slows down, actions such as the warming trend. Snow and ice, for example, reflect sunlight away from the earth. If warmer temperatures were to melt them, less sunlight would be reflected. More sunlight would then be absorbed by the earth's surface and would be turned into heat.

Experts disagree about how positive and negative feedbacks affect rising temperatures. But most scientists agree that the increase in greenhouse gases is causing global warming. They also conclude that if we keep spewing greenhouse gases into the atmosphere, the average temperature on our planet will go up by several degrees.

A GLOBAL SEESAW

Climate scientists use computer programs to forecast patterns such as global warming. Their studies must take into account natural processes, called feedbacks, that can boost or stop a predicted climate change.

Experts know, for instance, that about 80 percent of the light that strikes snow-covered areas is reflected back into the atmosphere, an action that cools the earth. When global warming melts snow and ice, the earth absorbs more light, and less heat is sent back toward outer space. This is an example of a positive feedback because the process promotes the warming trend.

Oceans give us an example of a negative feedback. Roughly 90 percent of the sunlight that hits the world's oceans is absorbed as heat. Some of the warmed water becomes water vapor and gathers in clouds. Because clouds reflect sunlight, more clouds might cause the earth to cool—an action that slows the warming trend.

It's important to remember that global climate patterns depend on many factors whose interaction is hard to pin down. Computers help experts make good guesses, but accuracy can only be judged by the pattern that actually takes place.

CHAPTER FOUR

CHANGED BY DEGREES

You might expect that a temperature increase of a few degrees wouldn't make much of a difference. But scientists think global warming poses one of the most serious dangers to life on our planet. A small upward shift in average global temperature could change our world dramatically.

WEIRD NEW WORLD

Any increase in average global temperature would result in hotter summers. Temperatures would reach 100° F (38° C) more often than ever before in cities such as Washington, D.C.; Lisbon, Portugal; and Ankara, Turkey. This change isn't just about making some of us sweat a little more.

Warmer temperatures could dislodge and then melt some of the huge, dense masses of ice around the North and South Poles. Much of the water from this melting ice would eventually end up in the oceans. The additional water would raise the sea level nearly three feet (one meter).

(Left) ***Shifts in global temperature could raise ocean levels and badly affect low-lying countries—such as Bangladesh in southern Asia.*** *(Above)* ***Sunbathers enjoy the beaches of Portugal, where temperatures could become uncomfortably hot if worldwide climate changes take place.***

Rising sea levels could destroy wilderness areas, such as the Florida Everglades, where many kinds of bird life currently make their homes.

Rising sea levels would flood parts of large coastal cities, such as Beijing, China, and New York, New York. Entire islands might end up underwater. One-fifth of Bangladesh, a low-lying nation in southern Asia, would no longer exist. The flooding would also cover huge areas of fertile soil at the mouths of large rivers, such as the Amazon in South America and the Nile in North Africa. The swampy Florida Everglades—an important breeding area for wildlife—would be lost forever. Throughout the world, millions of people would have to move to higher ground to escape rising seas.

Global warming might boost the number and destructive force of hurricanes and other storms in the Tropics, a region lying near the equator. The climate of the Tropics is warm year-round—and so are the surrounding ocean waters. Tropical hurricanes form when the temperature of the ocean water reaches at least 79° F (26° C). Global warming, as it raises the water's temperature, would create ideal conditions for more destructive hurricanes to take shape.

Scientists believe that global warming would also change the world's patterns of rain and snow. Some places would be drier than they are now, for instance. Others would be wetter. These changes could make survival difficult for many plants and animals, because the new climatic pattern wouldn't suit them.

Hurricanes and tornadoes could increase in number and force if the global warming trend continues.

Ukraine, a fertile country in eastern Europe, has some of the richest farmland in the world. But the nation's abundant harvests of grain might be affected by global warming.

Global climate changes would alter farming, too. Some areas would benefit from warmer temperatures. Japan, for example, might be able to grow four times the amount of rice it does now. But other areas would lose out. The rich croplands of Europe and central North America would become much drier. Farmers in these regions would probably be unable to grow the wheat and other grains that feed much of the world.

CAN WE ADAPT?

Humans and other living things would have little time to get used to the new

world. Rather than **adapt,** or change, to fit new conditions, plants and animals would have to "move" to survive.

Trees and other plants can't get up and walk away when conditions change. Instead, their seeds are naturally transplanted as their normal homes die off or become unlivable.

Birds and other animals—in search of their own new places to survive—unknowingly carry plant and tree seeds with them, perhaps as food or as nesting material. The seeds set down roots in a new environment, and eventually the plants reestablish themselves. But this is a slow process, and no one knows if it is fast enough to keep up with global warming.

Many living things, not just humans, would have trouble thriving on a planet with new climatic patterns. Birds that travel from season to season, such as scarlet tanagers** (above left), **might find conditions unlivable in their temporary homes. Scientific studies suggest that birch trees** (left) **would have to move farther north to survive.

Some species, or kinds of living things, would need more time to adapt. Others might be unable to make the adjustment at all. As a result, species—even entire **ecosystems**—could cease to exist altogether.

THE EVIDENCE

Will these changes actually happen if we keep adding greenhouse gases to our atmosphere? Although scientists fully understand the greenhouse effect, they are still gathering the scientific data that link the greenhouse effect to global climate change. But many people point out that the predictions seem to be coming true now.

Six of the seven warmest years of the last century and a half occurred during the 1980s, and 1990 was the warmest year ever recorded.

Mountain glaciers (slow-moving ice masses) have been melting and getting smaller over the past 100 years. Scientists have noted glacier shrinkages on every continent, from the Alps in Europe to the Andes in South America.

During the twentieth century, the level of the ocean has risen about 6 inches (15 centimeters).

A shrinking glacier in Canada's Banff National Park is retreating from the valley it once occupied.

A CLIMATIC BLAST

If global temperatures have risen every year, why was 1990 the hottest year on record? The answer may lie with Mount Pinatubo, a volcano in the Philippine Islands that erupted in June 1991. The blast spewed massive amounts of sulfur dioxide into the atmosphere. The sulfur mixed with water vapor in the air to create billions of droplets of sulfuric acid. The droplets eventually formed dense clouds.

The increased cloud cover acted as a shield, reflecting sunlight back toward outer space and cooling the earth slightly. Since the eruption, satellite readings have shown a 1° F (0.5° C) decrease in global temperature. Scientists also noted that the amount of atmospheric CO_2 and methane went down. Some researchers have suggested that the common cause for these events might be Mount Pinatubo.

A temporary halt in global warming is good news. But the blast also pumped ozone-destroying chemicals into the atmosphere, and these compounds sped up the loss of stratospheric ozone. The effects of Mount Pinatubo's eruption aren't permanent, however. Soon, we'll be struggling again with pre-eruption atmospheric conditions.

Mount Pinatubo erupted in 1991 after more than 600 years of inactivity.

Tropical storms and hurricanes seem to be fiercer than ever. The severe damage caused by Hurricanes Gilbert in 1988, Hugo in 1989, and Andrew in 1992 has led some people to believe that global warming already is affecting tropical climates.

Droughts (long periods without rain) have been more widespread in the past

Neighbors comfort one another after Hurricane Andrew, one of the most destructive storms ever to hit the United States.

Droughts have transformed the nations just south of the Sahara Desert on the African continent. Hardy animals and tough plants once thrived here, but now carcasses and sand dunes dot the landscape.

This macaroni penguin and other kinds of animals in the Antarctic region would have trouble surviving in a warmer climate.

several decades. Killing droughts hit many areas in Africa in the 1970s and 1980s. Other countries—including Australia, Bolivia, Italy, and the United States—also have fought hot, dry weather during recent years.

Winters in Antarctica have been warmer than usual for the last two decades. Scientists already are seeing an effect on Antarctica's animal life. Species that prefer the region's very cold winters are weakened by the warm temperature and therefore are less able to compete for food and living space.

Any or all of these changes could be just a case of bad luck. But with so many hints that global warming might be occurring—and with so many fearful results if it does—do we really want to wait and see?

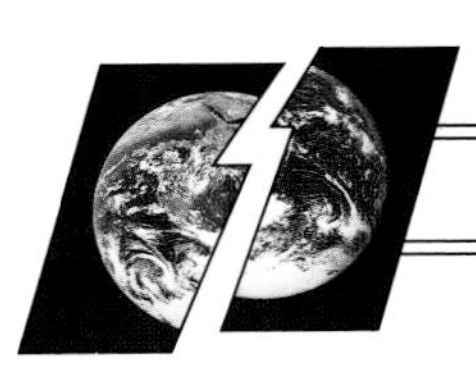

CHAPTER FIVE

THE KEY TO CHANGE

More than 25 years ago, the U.S. biologist Dr. Garrett Hardin told his fellow scientists a story. He asked them to imagine a village in which everyone grazes their cattle on the same grassy field.

Because the animals are all the villagers have, each person wants to raise as many cattle as possible. But the villagers also know that if too many animals share the land, they will eat up all the grass. The exposed soil, no longer protected by plant life, will be ruined and unable to grow more grass. Everyone's cows will starve.

Suppose one of the women in the village has the chance to buy another cow for her herd. What does she do? She knows that bringing in another animal could harm the land for everyone. Even if she decides not to add the cow, however, her neighbor may not choose to be so unselfish. The woman figures that the other villagers will increase their herds, and so she purchases the new animal.

Unfortunately, everyone in the village uses this same logic. The number of cattle soars, and eventually the grassy field is destroyed.

(Left) Cattle wander in a fertile plot in Mali, a sub-Saharan country where overgrazing (putting too many animals in a field to eat grass) has left some land unfit for farming.

GLOBAL TEAMWORK

You may wonder what the link is between this story and our planet's atmosphere. The countries of the world are the villagers, and the grassy field is the atmosphere we all share.

Dr. Hardin's tale helps us see what individual countries are up against when they try to solve the atmosphere's problems. Nobody *wants* to ruin the atmosphere. But some countries argue that their efforts to stop producing atmospheric gases won't make much difference unless there's a guarantee that other nations will stop, too.

This disagreement often comes up between two specific groups of nations. On one side are **industrialized nations.** These rich countries have built large-scale industries and burn lots of the world's fossil fuels. The United States, Canada, Germany, France, and Japan are among the world's industrialized nations.

On the other side are **developing nations.** These countries, although not wealthy, are setting up big industries, and they spend a lot of money on fossil fuels to run their factories. China, Brazil, Indonesia, Kenya, and India are some of the world's developing nations.

Industrialized nations manufacture most of the world's CFCs and contribute most of the additional greenhouse gases to the atmosphere. But CFCs and fossil fuels have

Workers in China, one of the world's most active developing nations, install air conditioners that contain CFCs. Most developing nations don't have or can't afford equipment to make their products environmentally safe, but the countries still want to expand manufacturing.

A cartoon shows the clash of interests between industrialized and developing nations.

enabled people in these nations to enjoy a high standard of living. Less-developed countries want to offer a more comfortable life to their citizens. These nations believe they need the same access to fuels and chemicals to achieve their goals.

Tom McMillan, Canada's Minister of the Environment, was among the politicians to sign the Montreal Protocol, the 1987 treaty that aimed to cut the use and production of CFCs.

But no solution will surface as long as each country looks out only for itself. Solutions depend on global cooperation.

SAVING THE OZONE

In the 1970s, proof of ozone loss was mounting. But most experts were not willing to blame the loss on CFCs in the atmosphere. Scientists took several more years to conclude that CFCs attacked ozone and that the world had produced enough CFCs to cause the problem. During those years of disbelief, more and more CFCs rose through the troposphere.

Nevertheless, a few countries took action. In 1978, the United States, Canada, Norway, and Sweden banned the use of CFCs in spray cans. This decision only put a dent in the problem, however. Large-scale global teamwork was needed, and the discovery of Antarctica's huge ozone hole sounded the grim wake-up call.

In 1987, when scientific evidence showed that we were destroying the protective shield above our heads, representatives of many nations gathered in Montreal, Quebec, in Canada. All participants agreed to freeze CFC use and production at 1986 levels and then to cut use and production in half by the year 2000. The agreement, called the Montreal Protocol, proved that countries could work together on a shared environmental issue.

Meanwhile, scientists found greater evidence that the problem was rapidly getting worse. Soon, the Montreal agreements weren't looking effective enough. In 1990, 93 countries got together in London, England, to review the protocol's goals. After much discussion, the delegates agreed that industrialized nations would completely stop making and using CFCs by 2000. In 1992, the deadline was moved up to 1996.

Keep in mind, though, that even if we halt all CFC production and use, we still won't have fully solved the ozone problem. Scientists know that CFCs survive a long time in the troposphere before reaching the stratosphere. The good news is that, because CFCs are people-made, we *can* stop their production. Already our efforts have given the ozone layer a chance to repair itself. But the ozone layer will still take from 50 to 100 years to get back to normal.

The October 1994 TOMS picture pointed out low ozone amounts over the South Pole as measured by Dobson Units (DUs). Note that the Dobson scale (right) goes from 475 down to 75 on this image. Previous scales ranged from 500 to 100. Starting in 1992, ozone levels dropped below 100, and the scale was adjusted to fit the new data.

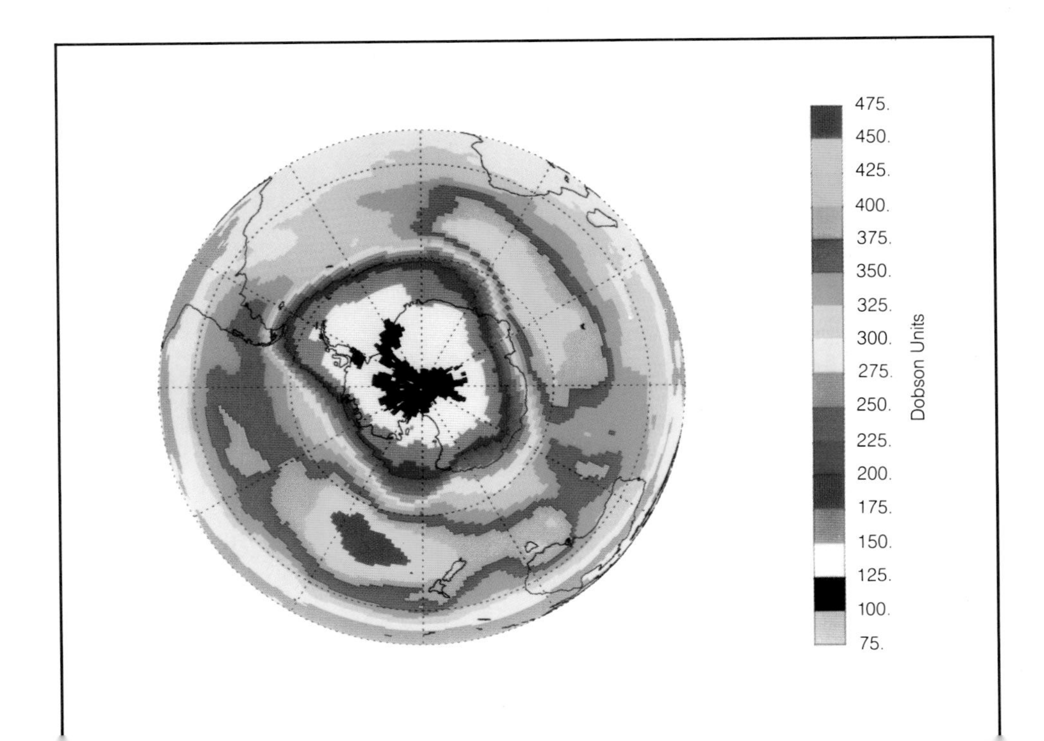

CHILLING OUT

When it comes to countries working together to fight global warming, the biggest hurdle is getting everyone to see the same problem. Most governments and scientists agree that the world must reduce greenhouse gases in the atmosphere. Yet there is disagreement on linking increases of greenhouse gases to global warming.

One solution is to cut back on all products, especially fossil fuels, that release greenhouse gases into the atmosphere. But people in industrialized countries drive a lot of cars, heat a lot of homes, and power a lot of factories with fossil fuels. Cutting CO_2 could mean major changes in the way people in these nations live.

At the same time, people in developing countries want their chance to own cars, to buy locally made factory goods, and to heat or cool their homes with electricity. These activities would add even more CO_2 to the atmosphere.

Some individual countries and private groups have chosen to do what they can to reduce the threat of global warming. The governments of Denmark, the Netherlands, Germany, and Sweden have developed programs to reshape all aspects of energy demand. These plans range from choosing less-harmful fuels to setting up new ways to use these fuels in homes and businesses.

Workers in India drive a gas-fueled tractor through an urban street. India wants to build a strong economy before limiting the amounts of greenhouse gases that can be put into the atmosphere.

SHARING THE WARMTH

Denmark and the Netherlands—two of Europe's most industrialized nations—are reshaping the way they use energy. Their aim is to reduce their countries' CO_2 emissions and fossil-fuel demands.

Laws passed by Denmark's government, for example, have encouraged the development of biogas plants. These facilities produce heat from naturally decaying materials, such as manure, straw, and wood chips.

Folks in the Netherlands have been using wind power—through their world-famous windmills—for centuries. Now the Dutch government is inviting energy companies to increase windpower capacity on a larger scale.

This cogeneration plant operates in the northeastern Netherlands.

Both countries are supporting cogeneration plants, where electricity and heat are produced and distributed at the same time. Facilities that burn fossil fuels to make electricity create heat in the process. But the plants usually release most of the heat as steam or hot water. Instead of wasting this heat, cogeneration plants trap it and send it by way of underground pipelines to nearby homes, businesses, and factories.

Dutch and Danish laws have spurred the manufacture of energy-efficient dishwashers, refrigerators, and other appliances. These new machines cost about the same, use about half the energy, and can run on cogenerated power.

At a trash-burning plant in Michigan, crews install modern equipment to control emissions (releases) of greenhouse gases. This technology is not always affordable or available to factories in developing countries.

In 1993, the U.S. government introduced the Climate Change Action Plan. The plan encourages cooperation among governmental groups, private businesses, and local communities throughout the country. Local energy companies, for example, are checking the efficiency of household heating systems and are suggesting energy-saving improvements.

INTERNATIONAL EFFORTS

Cooperation among governments is gradually growing, too. The United Nations Environment Programme (UNEP) and the World Meteorological Organization created the Intergovernmental Panel on Climate Change (IPCC) in 1988. The group's job was to put together a worldwide plan for dealing with global warming.

One of the IPCC's reports concludes not only that the amount of greenhouse gases is multiplying but also that the increase contributes to global warming. Some members of the group have even recommended that countries with seacoasts start planning for a rise in sea level.

So far, however, countries have been unable to agree on specific limits to the production of greenhouse gases. The conflict between industrialized and develop-

In 1992, in Rio de Janeiro, Brazil, members of the United Nations Conference on Environment and Development—also known as the Earth Summit—discussed global warming. By 1994, many nations had agreed to the voluntary goal of reducing greenhouse gas emissions.

ing nations is a big part of this debate. The United States and Japan, for example, are responsible for adding two-thirds of all greenhouse gases to the atmosphere. Developing nations ask why they should limit their much smaller output. After all, rich nations—even after cutbacks—still contribute a much higher percentage of these gases.

Environmentalists hoped that limits on greenhouse gas production would be set in 1992. In that year, the United Nations sponsored the Earth Summit in Rio de Janeiro, Brazil. Global warming was a key topic among the summit's thousands of participants.

Governments represented at the summit did agree to reduce CO_2 production to 1990 levels by the end of the twentieth century. But the document they signed made this a voluntary goal and did not set specific timetables to achieve it.

Many people believe that the conference participants didn't go far enough. Nevertheless, by January 1994 the Climate Convention—the document signed in Rio—had been formally and legally adopted by more than 50 of the world's governments.

CHAPTER SIX

WHAT CAN WE DO?

Solving global threats to the atmosphere is not just the job of scientists and politicians. Solutions also involve personal actions. And individual actions multiplied by billions of people can really make a difference! Here are steps all of us can take to reduce our impact on the atmosphere.

CONSERVE ENERGY. People in industrialized nations use a lot of electricity and heat, much of which comes from facilities that run on fossil fuels. Turn off lights, TVs, radios, and other users of electricity when you don't need them. When you get cold, put on a sweater instead of turning up the thermostat.

(Left) ***People who want to preserve the world's environment walked together on Earth Day in Oregon.*** *(Right)* ***At a rally in Ontario, Canada, a girl expressed her concern for the future.***

REDUCE, REUSE, RECYCLE, DO WITHOUT. The amount of garbage we make and how we throw it away can have a long-term effect on the atmosphere. When we buy less, we have less to feed methane-producing bacteria in landfills. When we reuse or recycle, we decrease the need for fossil fuels to run factories that make new products. When we do without, we take action against the huge amounts of the earth's resources that we use. If your school doesn't yet have a recycling program, ask a teacher or parent to help you start one.

BE A WISE SHOPPER. Avoid buying products that have too much packaging. Support companies that make their goods—or at least their packaging—from recycled materials. This information will usually be marked on the package.

USE OTHER TRANSPORTATION BESIDES CARS. Exhaust from cars adds tons of CO_2 to our atmosphere, worsening the greenhouse effect and adding to global warming. Mass-transit systems—buses, subways, and trains—are good alternatives to cars because they carry many people at once. And bicycles and walking are the cleanest and cheapest ways to get from one place to another.

PLANT AND PRESERVE TREES. Trees take in CO_2 as they build roots, stems, leaves, and branches. If more CO_2

Recycling as many waste products as possible lessens the amount of garbage we put in landfills, where methane can form.

is used by trees, smaller amounts of the gas will build up in the atmosphere. Trees are also natural coolants, providing shade to people and buildings. By having more trees around, we can keep cool without electric fans and without air conditioners, which often contain CFCs.

ACTING LOCALLY

On a clear day, residents of Portland, Oregon, can see Mount Hood. To make sure the view will always be there, the city council passed the Carbon Dioxide Reduction Strategy to lower urban CO_2. Information packets help Portland's families learn about their energy needs. The handouts also encourage households to buy more energy-efficient appliances, to recycle more waste, and to use an alternative to cars at least once a week. The local government has purchased vehicles that run on alternative fuels, has promoted tree planting and tree maintenance, and has supported the development of cogeneration and biogas facilities.

ONE TREE AT A TIME

American Forests, a U.S. conservation organization, is spearheading two tree-planting projects—Global ReLeaf and Cool Communities. Global ReLeaf helps people throughout the world raise and care for trees. So far, volunteers in the program have planted 2.5 million seedlings in the United States alone and have advised planters in Kenya, South Africa, Ukraine, Bulgaria, Canada, Ecuador, and Costa Rica.

Cool Communities is working to reduce the heat in U.S. cities, where huge amounts of fossil fuels are burned. Surveys by American Forests reveal that most cities are uprooting more trees than they are planting and that trees in crowded downtown areas last for only 13 years. Cool Communities hopes to increase the number of new trees and also to better maintain existing trees.

Trees take in CO_2 as they grow, and they also provide shade. Shaded buildings require less air conditioning, which uses ozone-destroying CFCs and runs on electricity produced by fossil fuels. American Forests estimates that three mature shade trees placed on the southeast and southwest sides of a house can cut air-conditioning costs by 50 percent. And each year, just one growing tree absorbs 26 pounds (12 kilograms) of CO_2.

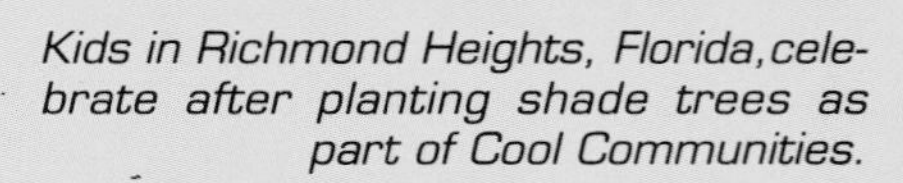

Kids in Richmond Heights, Florida, celebrate after planting shade trees as part of Cool Communities.

Friends** (right) **get together to write letters about protecting the environment of Antarctica. Taking part in public gatherings** (below right) **helps educate those around us, including our families, our friends, and our local leaders.

WRITE LETTERS. Put pressure on the governments, agencies, and people who are making decisions about repairing the atmosphere. Tell these people that you care about the ozone layer and global warming. Write to your local newspaper or school newsletter, too. If they print your message, you could help change the thinking of hundreds or thousands of other people.

EDUCATE THE PEOPLE AROUND YOU. Help your friends and family take a new look at their role in the global environment. Show them how their actions, especially driving cars, contribute to the atmosphere's problems. Let these people also know, though, that there are lots of ways we can help to heal the atmosphere. Then encourage your friends and family to join you in these efforts.

ORGANIZATIONS

AIR AND WASTE MANAGEMENT ASSOCIATION
1 Gateway Center, Third Floor
Pittsburgh, Pennsylvania 15222

AMERICAN FORESTS
Global ReLeaf
Post Office Box 2000
Washington, D.C. 20013

CENTER FOR ENVIRONMENTAL STUDY
Grand Rapids Community College
143 Bostwick NE
Grand Rapids, Michigan 49503

CONCERN, INC.
1794 Columbia Road NW
Washington, D.C. 20009

ENVIRONMENTAL DEFENSE FUND, INC.
257 Park Avenue South
New York, New York 10010

GREENHOUSE ACTION
Post Office Box 68218
Seattle, Washington 98168

NATIONAL ARBOR DAY FOUNDATION
Conservation Trees
100 Arbor Avenue
Nebraska City, NE 68410

UNITED STATES COMMITTEE FOR THE UNITED NATIONS ENVIRONMENT PROGRAMME
2013 Q Street NW
Washington, D.C. 20009

Photo Acknowledgments

Photographs are used courtesy of: Michael Mogil, p. 1; NASA, pp. 4, 23, 25, 55, 69 (right); Kaisei-sha Publishing Co., Japan, pp. 6, 28; Dr. A. A. M. van der Heyden, p. 7; NSF, pp. 9, 24, 26 (right); NCAR, p. 11 (top); United Airlines, p. 11 (bottom); GA Dept. of Industry, Trade, and Tourism, p. 12; © John Kreul, p. 13; Robert E. Olson, p. 15; © Richard B. Levine, p. 19; DuPont, p. 20 (left); IPS, p. 20 (right); UC-Irvine, p. 22; Australian Consulate-General, p. 26 (left); Joe H. Sullivan, p. 27; Leonard Soroka, p. 30; Herbert Fristedt, p. 32; Janda Thompson, p. 34; AID, p. 35 (right); Visuals Unlimited: George Loun, p. 36, N. Pecnik, p. 39, Martin G. Miller, p. 46, G. Prance, p. 49; Rick Hansen / MN Dept. of Agriculture, p. 37; CIDA Photo / Roger Lemoyne, p. 40; Dr. Roma Hoff, p. 41; Ned Skubic, p. 42; MN DNR, p. 43; Jeff Greenberg, p. 44; Perry J. Reynolds, p. 45 (top); Olive Glasgow, p. 45 (bottom); U.S. Air Force / SRA Paul Davis, p. 47; Jim Virga / The Sun-Sentinel, p. 48 (top); FAO, p. 48 (bottom); The Hutchison Library, p. 50; OPIC, p. 52; Scott Willis, San Jose Mercury News, p. 53; Environment Canada, p. 54; M. Bryan Ginsberg, pp. 56, 60; Ministry of the Environment, the Netherlands, p. 57; © Jim West, pp. 58, 61, 65 (bottom); UN Photo, p. 59; Jerry Boucher, p. 62; © Larry Geddis, p, 63; American Forests / © Daniel C. Smith; Patricia Drentea, p. 65 (top); Inter-American Development Bank, p. 67; Husky Oil Corp., p. 68; Mitsubishi, p. 69 (left); James H. Carmichael, p. 70. Maps and charts: Laura Westlund, pp. 8 (top and bottom), 31, 33; Darren Erickson, p. 16; Bryan Liedahl, pp. 10, 18, 21; Jean Matheny, p. 35.

Front Cover: NASA
Back Cover: (left) Jeff Greenberg; (right) © Shmuel Thaler

adapt: to change to fit new conditions.

air pollution: the presence of high amounts of impurities in the air that present danger to health and comfort.

anaerobic bacteria (aa-nuh-ROH-bik bak-TEER-ih-yah): tiny life-forms that break down dead matter in an environment that contains no oxygen.

atom: one of the smallest parts of a chemical element, such as carbon or oxygen.

Workers package fish at a food-processing plant in Peru, a developing nation in South America.

carbon dioxide (CO_2): a gas that is naturally found in the atmosphere in small amounts. We add CO_2 to the air by burning wood, by running vehicles and factories on gasoline or coal, and even by breathing out.

chloro-fluoro-carbons (CFCs): chemicals made up of chlorine, fluorine, and carbon that have many industrial uses. CFCs are used in making some foam packaging, cleaning fluids, and air coolants.

core sample: a long, narrow column of material obtained by drilling deep into the earth with a hollow tube. In ice-covered areas, the drill gathers samples from many different layers of ice, some of which formed millions of years ago.

developing nation: a country that is just beginning to develop its natural resources and that has few industries.

ecosystem: a complex community of living and nonliving things that exists as a balanced unit in nature.

food web: a group of plants and animals, each of which is a source of food for the next, or larger, member in the web.

At refineries, petroleum, one of the world's most valued fossil fuels, is made into gas, oil, and other products.

fossil fuel: a substance, such as coal or petroleum, that slowly developed from the remains of living things.

global warming: an increase in the earth's average temperature caused by the strengthening of the greenhouse effect.

greenhouse effect: the result of the sun's heat becoming trapped in the earth's atmosphere by gases in the same way that glass traps heat in a greenhouse.

greenhouse gas: a gas—such as carbon dioxide, methane, or nitrous oxide—that traps heat in the atmosphere.

industrialized nation: a country that has developed its natural resources and that has set up a wide variety of industries.

infrared radiation: an invisible form of energy from the sun that we feel as heat.

mesosphere: the third layer of the earth's atmosphere.

molecule: the smallest combination of atoms that will form a given chemical compound, such as carbon dioxide or ozone.

negative feedback: a natural response that reduces or stops a certain chemical or biological process.

ozone: a naturally occurring gas found mostly in the stratosphere.

Laborers (left) in Japan, a highly industrialized country, assemble a new automobile. A researcher (below) watches instruments on board a plane to measure ozone levels above Antarctica.

ozone layer: the thin sheet of ozone gas in the stratosphere that shields the earth by absorbing the sun's harmful rays.

photosynthesis (fote-oh-SIN-thuh-suss): the chemical process by which plants make their own food. The process uses carbon dioxide, water, and sunlight.

phytoplankton (fight-oh-PLANK-tin): tiny plants that are moved in water by waves or currents. These plants are the main food source for some members of the food web.

positive feedback: a natural response that increases a certain chemical or biological process.

stratosphere: the second layer of the earth's atmosphere.

thermosphere: the fourth and outermost layer of the earth's atmosphere.

troposphere: the first and innermost layer of the earth's atmosphere.

ultraviolet B (UV-B) ray: a form of energy from the sun that humans cannot see directly. UV-B rays are largely absorbed by the ozone layer.

visible light: a form of energy from the sun that humans can see.

During photosynthesis, green plants use CO_2, water, and energy from the sun to feed themselves.

INDEX